I0471172

Canadian Research on Pedestrian

Safety

PUBLICATION NO. FHWA-RD-99-090

DECEMBER 1999

U.S. Department of Transportation

Federal Highway Administration

Research, Development, and Technology
Turner-Fairbank Highway Research Center
6300Georgetown Pike
McLean, VA 22101-2296

FOREWORD

Creating improved safety and access for pedestrians requires providing safe places for people to walk, as well as implementing traffic control and design measures which allow for safer street crossings. A study entitled "Evaluation of Pedestrian Facilities" involved evaluating various types of pedestrian facilities and traffic control devices, including pedestrian crossing signs, marked versus unmarked crosswalks, countdown pedestrian signals, illuminated pushbuttons, automatic pedestrian detectors, and traffic calming devices such as curb extensions and raised crosswalks. The study provided recommendations for adding sidewalks to new and existing streets and for using marked crosswalks for uncontrolled locations. The "Evaluation of Pedestrian Facilities" also included synthesis reports of both domestic and international pedestrian safety research. There are five international pedestrian safety synthesis reports; this document compiles the most relevant research from Canada.

This synthesis report should be of interest to State and local pedestrian and bicycle coordinators, transportation engineers, planners, and researchers involved in the safety and design of pedestrian facilities within the highway environment.

Michael F. Trentacoste
Director, Office of Safety
Research and Development

NOTICE

This document is disseminated under the sponsorship of the Department of Transportation in the interest of information exchange. The U.S. Government assumes no liability for its contents or use thereof. This report does not constitute a standard, specification, or regulation.

The U.S. Government does not endorse products or manufacturers. Trade and manufacturer's names appear in this report only because they are considered essential to the object of the document.

1. Report No. FHWA-RD-99-090	2. Government Accession No.	3. Recipient's Catalog No.

4. Title and Subtitle	5. Report Date
Canadian Research on Pedestrian Safety	
	6. Performing Organization Code

7. Author(s)	8. Performing Organization Report No.
Ron Van Houten and J.E. Louis Malenfant	

9. Performing Organization Name and Address	10. Work Unit No. (TRAIS)
Mount Saint Vincent Univ. and Center for Educ. & Res. in Sfty University of North Carolina Highway Safety Research Center 730 Airport Rd, CB #3430 Chapel Hill, NC 27599-3430	11. Contract or Grant No. DTFH61-92-C-00138

12. Sponsoring Agency Name and Address	13. Type of Report and Period Covered
Federal Highway Administration Turner-Fairbanks Highway Research Center 6300 Georgetown Pike McLean, VA 22101-2296	14. Sponsoring Agency Code

15. Supplementary Notes

Prime Contractor: University of North Carolina Highway Safety Research Center
FHWA COTR: Carol Tan Esse

16. Abstract

This report was one in a series of pedestrian safety synthesis reports prepared for the Federal Highway Administration (FHWA) to document pedestrian safety in other countries. Reports are also available for:

United Kingdom (FHWA-RD-99-089)
Sweden (FHWA-RD-99-091)
Netherlands (FHWA-RD-99-092)
Australia (FHWA-RD-99-093)

This review reports research in six areas of pedestrian safety:

1) Interventions to prompt pedestrians to watch for turning vehicles.
2) Improving pedestrian signals for better indication of clearance interval.
3) Use of pedestrian-activated beacons at uncontrolled crossings.
4) Use of advance stop lines.
5) Increasing conspicuity of crosswalks.
6) Use of multiple interventions to increase motorist yielding to pedestrians.

Research results are presented and a comprehensive list of references is provided.

17. Key Words: pedestrians, flashing beacons, crosswalks, pedestrian signals, moving eyes display, pedestrian signs, advance stop lines	18. Distribution Statement

19. Security Classif. (of this report) Unclassified	20. Security Classif. (of this page) Unclassified	21. No. of Pages	22. Price

Form DOT F 1700.7 (8-72) Reproduction of form and completed page is authorized

SI* (MODERN METRIC) CONVERSION FACTORS

APPROXIMATE CONVERSIONS TO SI UNITS

Symbol	When You Know	Multiply by	To Find	Symbol
LENGTH				
in	inches	25.4	millimeters	mm
ft	feet	0.305	meters	m
yd	yards	0.914	meters	m
mi	miles	1.61	kilometers	km
AREA				
in^2	square inches	645.2	square millimeters	mm^2
ft^2	square feet	0.093	square meters	m^2
yd^2	square yards	0.836	square meters	m^2
ac	acres	0.405	hectares	ha
mi^2	square miles	2.59	square kilometers	km^2
VOLUME				
fl oz	fluid ounces	29.57	milliliters	mL
gal	gallons	3.785	liters	L
ft^3	cubic feet	0.028	cubic meters	m^3
yd^3	cubic yards	0.765	cubic meters	m^3

NOTE: Volumes greater than 1000 l shall be shown in m^3.

Symbol	When You Know	Multiply by	To Find	Symbol
MASS				
oz	ounces	28.35	grams	g
lb	pounds	0.454	kilograms	kg
T	short tons (2000 lb)	0.907	megagrams (or "metric ton")	Mg (or "t")
TEMPERATURE				
°F	Fahrenheit temperature	5(F-32)/9 or (F-32)/1.8	Celcius temperature	°C
ILLUMINATION				
fc	foot-candles	10.76	lux	lx
fl	foot-Lamberts	3.426	candela/m^2	cd/m^2
FORCE and PRESSURE or STRESS				
lbf	poundforce	4.45	newtons	N
lbf/in^2	poundforce per square inch	6.89	kilopascals	kPa

APPROXIMATE CONVERSIONS FROM SI UNITS

Symbol	When You Know	Multiply by	To Find	Symbol
LENGTH				
mm	millimeters	0.039	inches	in
m	meters	3.28	feet	ft
m	meters	1.09	yards	yd
km	kilometers	0.621	miles	mi
AREA				
mm^2	square millimeters	0.0016	square inches	in^2
m^2	square meters	10.764	square feet	ft^2
m^2	square meters	1.195	square yards	yd^2
ha	hectares	2.47	acres	ac
km^2	square kilometers	0.386	square miles	mi^2
VOLUME				
mL	milliliters	0.034	fluid ounces	fl oz
L	liters	0.264	gallons	gal
m^3	cubic meters	35.71	cubic feet	ft^3
m^3	cubic meters	1.307	cubic yards	yd^3
MASS				
g	grams	0.035	ounces	oz
kg	kilograms	2.202	pounds	lb
Mg (or "t")	megagrams (or "metric ton")	1.103	short tons (2000 lb)	T
TEMPERATURE				
°C	Celcius temperature	1.8C+32	Fahrenheit temperature	°F
ILLUMINATION				
lx	lux	0.0929	foot-candles	fc
cd/m^2	candela/m^2	0.2919	foot-Lamberts	fl
FORCE and PRESSURE or STRESS				
N	newtons	0.225	poundforce	lbf
kPa	kilopascals	0.145	poundforce per square inch	lbf/in^2

*SI is the symbol for the International System of Units. Appropriate rounding should be made to comply with Section 4 of ASTM E380.

(Revised September 1993)

TABLE OF CONTENTS

1. Introduction

Canadian research in the area of pedestrian safety has focused on six areas of investigation:

1. Interventions to prompt pedestrians to look for turning vehicles when crossing at signalized crosswalks, including modification of the pedestrian signal head.

2. Modification of pedestrian signals to increase the clarity of the indication for the clearance interval.

3. The use of pedestrian activated flashing beacons at midblock crosswalks and at crosswalks on major roads at intersections not controlled by traffic signals.

4. The use of advance stop lines to increase the safety of pedestrians at crosswalks.

5. Research on interventions to increase the conspicuity of crosswalks.

6. The use of multifaceted programs that focus on engineering, enforcement, and education (the three E's) to increase yielding to pedestrians in crosswalks.

This paper will review research carried out in these six areas.

2. Use of Prompts to Reduce Threats Posed by Turning Vehicles

The percentage of pedestrian crashes that occurs at intersections is particularly high in urban areas in Canada. For example, an analysis of motor vehicle collisions with pedestrians in the province of Ontario found the majority of injury crashes occurred at intersections (Lane, McClaffery, & Nowak, 1996). This parallels experience in the United States where one fifth of motor vehicle crashes at signalized intersections involve a turning vehicle striking a pedestrian (Robertson & Carter, 1984).

Habib (1980) documented an over representation of left-turning vehicles in pedestrian crashes at intersections finding left-turning vehicles were about four times as hazardous as through movements. One reason why left-turning vehicles may be over represented in serious pedestrian crashes is the larger turning radius of left-turning vehicles enables them to travel at a higher velocity. Quaye, Leden, and Hauer (1993) examining crashes in Hamilton, Ontario, found that the probability of a pedestrian collision with a left-turning vehicle varied as a function of traffic volume and type of left-turn signal phasing. Quaye et al. speculated that these types of crashes may be related to the low level of observing behavior exhibited by motorists and pedestrians using crosswalks with traffic and pedestrian signals. Lord (1996) obtained similar results when he evaluated the same intersections used in Quaye et al's. study, and he also found a high correlation between pedestrian motor vehicle conflicts and crash history at these sites.

Van Houten, Retting, Malenfant, and Van Houten (1995) using data collected in the Halifax Regional Municipality in Nova Scotia found that serious motor vehicle/pedestrian conflicts occur at a moderate frequency for vehicles turning right on green and at a high frequency for vehicles turning left on green. These findings are in accord with the data published by others showing that left-turning vehicles are over represented in crashes at crosswalks.

When Van Houten and Malenfant examined pedestrian "observing" behavior across the relative location of threats, they found the percentage of pedestrians looking for turning vehicles was highest for vehicles starting their turn ahead of the pedestrian, lower for vehicles starting their turn beside the pedestrian, and lowest for vehicles starting their turn behind the pedestrian. These data showed that there is a strong inverse relationship between the occurrence of motor vehicle/ pedestrian conflicts and the level of pedestrian observing behavior. Jennings, Burki, and Onstine (1977) also reported that pedestrians tended to search more for potential threats while crossing during the "DON'T WALK" phase then while crossing during the "WALK" phase. It has also been reported that pedestrian-search-and-detection failures are the most common cause of pedestrian/motor vehicle crashes after inappropriate crossing (Shinar, 1978).

Zegeer, Cynecki, and Opiela (1984) found that "PEDESTRIANS WATCH FOR TURNING VEHICLES" signs reduced motor vehicle/pedestrian conflicts at a number of signalized crosswalks. Retting, Van Houten, Malenfant, Van Houten, and Farmer (1996) found that signs requesting pedestrians to look for turning vehicles erected next to the pedestrian signal head, or a similar message painted in the crosswalk, produced enduring increases in the percentage of pedestrians looking for all threats and almost eliminated conflicts between pedestrians and turning vehicles. Similar increases in observing behavior and reductions in conflicts were also produced using a digitally recorded verbal message played at the start of the WALK phase prompting pedestrians to look for turning vehicles (Van Houten, Malenfant, Van Houten, and Retting,1998) . The reductions in conflict frequency reported in these studies take on considerable significance given the high correlation between the type of conflicts scored in these studies and the incidence of pedestrian crashes (Lord,1996).

The use of paint, signs, and audible messages has been shown to be effective in prompting pedestrians to look for turning vehicles, thus reducing conflicts therewith. Some of these effects persisted for up to three years, though wide-scale implementation of these prompts would prove costly. A more economical way to increase pedestrians' observing behavior would be to incorporate the prompt as part of the WALK indication. Zegeer et al. (1984) evaluated a "WALK WITH CARE" signal indication as part of an experimental three-section signal head. They found that the "WALK WITH CARE" display produced a marked reduction in conflicts between pedestrians and right- and left-turning vehicles at four test intersections. One disadvantage of the experimental head used by Zegeer et al. (1984) is that it employed a written message rather than an international symbol and hence may not be understood by tourists and others who may not speak English. Furthermore, research also indicated that the WALK and DON'T WALK symbols are more effective than the written message even when they are equally understood (Robertson, 1977) and therefore, it might expected that a symbolic message prompting pedestrians to look for turning vehicles might be more effective than a written message.

Van Houten, Van Houten, Malenfant, and Retting (1998) evaluated the use of symbolic indication prompting pedestrians to look for turning vehicles. It consisted of adding animated eyes that searched from side to side to the "WALK" indication at two signalized intersections. The length of the "WALK" indication was 7 seconds on the main street, 30 seconds on one of the secondary streets, and

40 seconds on the other secondary street. Observers scored the looking behavior of pedestrians and pedestrian/motor vehicle conflicts on weekdays between 8:30 a.m. and 5 p.m.

Pedestrians at each of the crosswalks had to cross three potential turning vehicle paths. Figure 1 shows the three possible conflict points for one of the four crosswalks. A pedestrian crossing in a clockwise direction would first encounter a potential conflict with a vehicle turning right on red at C, next the pedestrian would encounter a potential threat from a vehicle turning left on green at B, and finally the pedestrian would encounter a potential threat with a vehicle turning right on green at A. A pedestrian crossing in the counterclockwise direction would encounter these three threats in the opposite order. Pedestrians were scored for checking these three threats in the order they were incurred crossing the intersection. To be scored as checking a particular threat, the pedestrian had to orient his or her head toward the direction the vehicle would be coming from prior to and within 3 seconds of entering the potential vehicle path. A motor vehicle/pedestrian conflict was scored if the turning vehicle had to engage in abrupt braking, or had to swerve in order to avoid striking the pedestrian who was being observed, or if the pedestrian had to take sudden evasive action to avoid being struck.

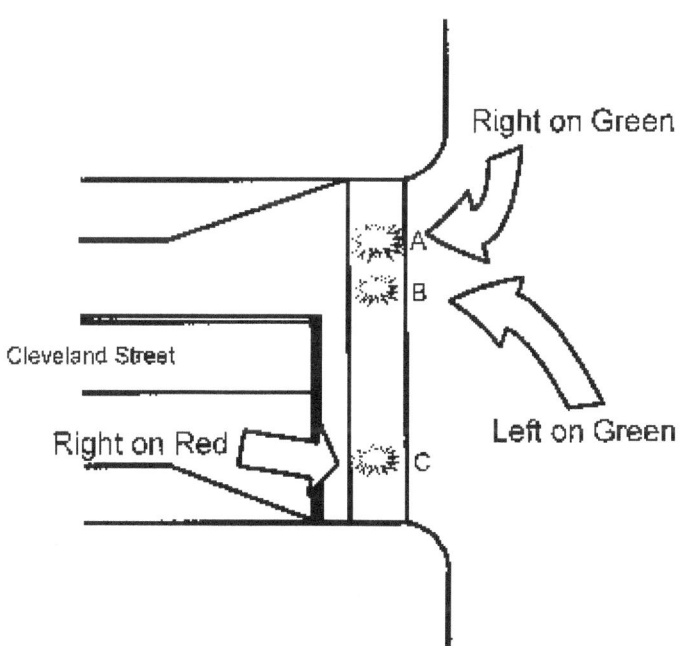

Figure 1. The three conflict paths that a pedestrian has to cross when crossing a street at the junction of two streets with two-way traffic and no turn restrictions.

The EYES display consisted of two blue eyes with blue eyeballs that scanned left and right at a rate of one cycle per second. This pictographic symbol was constructed from blue (460 nm) LEDs with an 8-degree field of view so that it would be primarily visible to pedestrians. Each eye was 127 mm (5 in) wide and 68.58 mm (2.7 in) high. The two eyes were separated by 57.15 mm (2 1/4 in). The WALK indication used was an outline of a walking person on a black background constructed from blue LEDs with an 8-degree field of view. The DON'T WALK indication used was a steadily illuminated outline of an upraised hand illuminated by orange (615 nm) LEDs with an 8-degree field of view on a black background.

Laboratory testing of the device with 100 English and 100 French (the two official languages used in Canada) speaking university students was conducted prior to beginning the field research to determine how they interpreted the EYES display in the context of a pedestrian signal head. All subjects identified the symbol as representing eyes and indicated that the purpose of the signal was to remind them to look for traffic. These results indicate that the meaning of the symbol is clear; it does not require special educational efforts to understand it; and it would be a good choice for international application. A photograph of the signal head showing the WALK indication with the EYES display is presented in figure 2.

Figure 2. A photograph showing the experimental head with the WALK indication and the EYES display illuminated together.

Van Houten et al. (1998) employed a multiple baseline design in this study. In a multiple baseline design, the treatment is introduced at a different point in time on each of the streets to control other factors that may have changed along with the introduction of the experimental intervention. In this study the experimenters compared the traditional incandescent pedestrian head with the use of the LED pedestrian head but without the use of the animated eyes display. Figure 3 is an event diagram showing the timing of each of the experimental conditions that included the use of animated eyes. After collecting baseline data at both intersections, the experimental pedestrian heads were first introduced at the one intersection without the EYES display to control any effects the novel LED pedestrian head might have on pedestrian behavior. The use of the LED display had no effect on pedestrians observing behavior or pedestrian/motor vehicle conflicts.

Next the EYES display was added at the first intersection so it came on alone during the first 2.5 seconds of the WALK interval and then was replaced by the standard pedestrian symbol for the duration of the WALK interval. This condition lead to a marked increase in pedestrians' observing behavior and a marked reduction in pedestrian/motor vehicle conflicts for pedestrians leaving early during the WALK interval at both sites from 2.7 conflicts per 100 crossing to 0.5 conflicts per 100 crossings. However, most pedestrians would not begin to cross until the standard "WALK" indication appeared. This resulting reduction in available WALK time associated with this timing sequence could be a disadvantage at intersections with short "WALK" indications.

The second presentation method evaluated was the simultaneous use of the EYES display and the standard walking man symbol for the first 2.5 seconds of the WALK interval followed by the termination of the EYES display for the remaining WALK time. This presentation method produced the same benefits as the sequential presentation method, and pedestrians did not lose any available WALK time. During the final condition, the EYES and man display were presented simultaneously for the initial 2.5 seconds, and then the EYES display switched off and reappeared for 2.5 every 9.5 seconds to prompt pedestrians who did not begin to cross at the start of the WALK interval to watch for turning vehicles. This presentation method maintained high levels of observing behavior and near zero levels of pedestrian/motor vehicle conflicts that persisted for pedestrians that left the curb during the entire WALK interval. These effects were all found to be statistically significant.

During the final 2 days of the study, a research assistant surveyed 100 pedestrians crossing at the experimental crosswalk. Pedestrians were asked: What they thought the new animated signal at the top of the pedestrian head was; what they thought of the new signal; and whether they would like to see this signal implemented elsewhere. The results of the survey indicated that all of the respondents identified the EYES display as eyes and that they understood the purpose was to tell them to look. Peoples' reaction to the signal was very positive and enthusiastic, and most of the respondents indicated that they would like to see the EYES display implemented elsewhere.

These results support and extend the findings of Zegeer et al. (1984) that modifying the pedestrian head to prompt pedestrians to take care while crossing the street is highly effective in

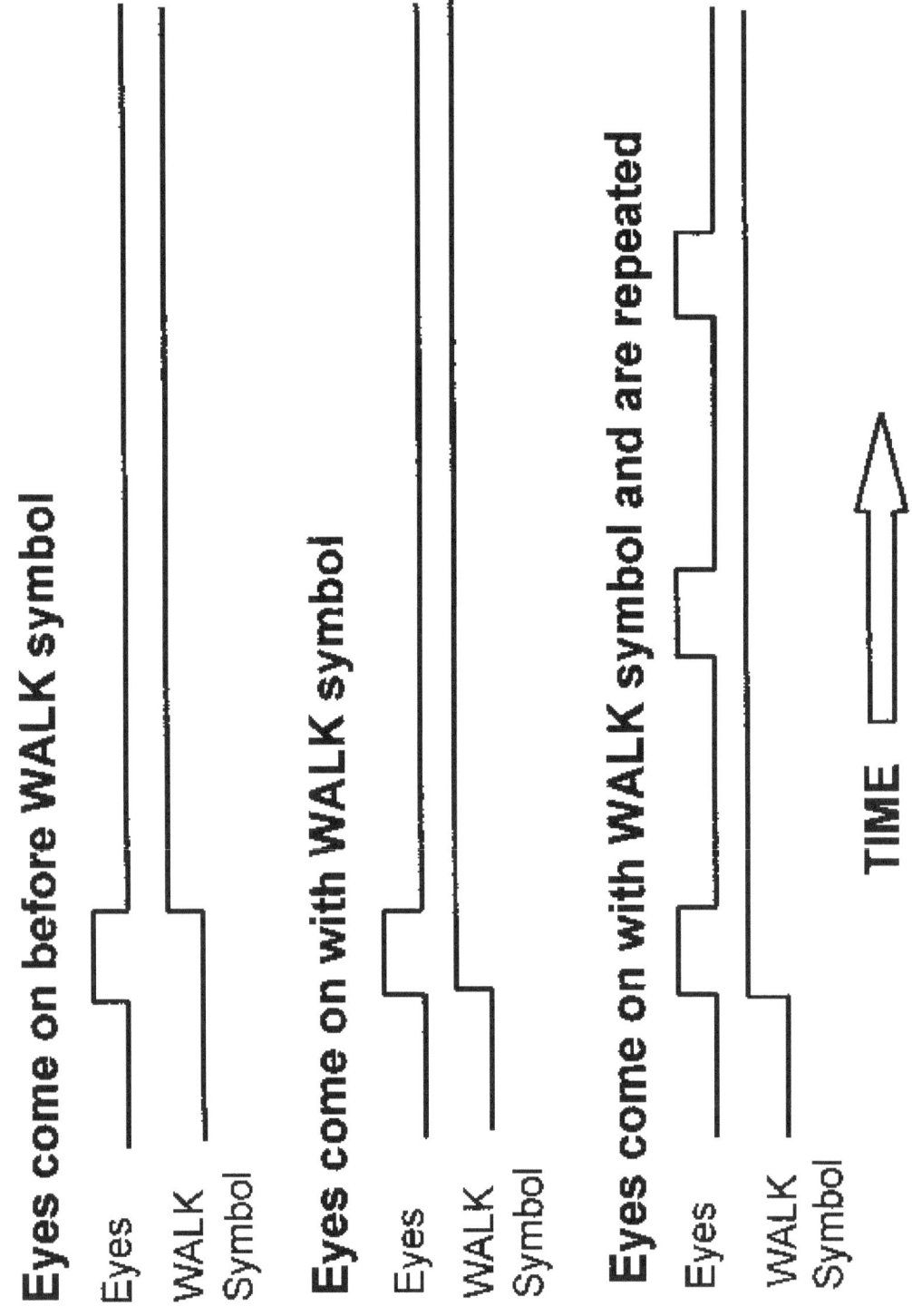

Figure 3. An event diagram showing the timing used in each of the experimental conditions.

Eyes come on before WALK symbol

Eyes

WALK
Symbol

Eyes come on with WALK symbol

Eyes

WALK
Symbol

Eyes come on with WALK symbol and are repeated

Eyes

WALK
Symbol

TIME

reducing pedestrian/motor vehicle conflicts, and provides additional data showing that the mechanism responsible for conflict reduction is increased pedestrian-observing behavior.

3. Research to Increase the Clarity of Pedestrian Clearance Interval

Another concern that has been experimentally examined by Canadian researchers is the poor level of compliance and understanding associated with the flashing hand indication for the pedestrian clearance interval. In one study Gourvil, Pellerin, and Hassan (1994) evaluated whether the use of a tricolored pedestrian heads would be better understood by pedestrians than the standard two-colored pedestrian head (white silhouette of a pedestrian and an orange hand) and therefore increase the safety of pedestrians at crosswalks. The tricolored pedestrians head used in this study consisted of one symbol, a silhouette of a walking pedestrian, combined with the use of a green, yellow, and red pedestrian head in a vertical configuration similar to that used with the standard green, yellow, and red traffic signals. A green silhouette light of a walking pedestrian was used for the "WALK" phase, a yellow silhouetted light was used for the "DON'T BEGIN TO CROSS" phase (to replace the flashing orange hand), and a red silhouetted light was used for the "DON'T WALK" phase (to replace the orange hand).

Eight intersections in six Quebec municipalities were selected for this study. The tricolored pedestrian heads were installed, and an 11-question survey was used to interview 1,917 pedestrians before and after the new pedestrian heads were installed. Pedestrians behavior at these crosswalks was also observed before and after the new signals were installed to determine the level of compliance to the standard and tricolored pedestrian heads.

The results of the pedestrian survey indicated that the tricolored pedestrian head was better understood than the standard pedestrian head. There was no difference in pedestrian understanding between the standard pedestrian heads and the tricolored heads for the "WALK" and "DON'T WALK" indications, however there was an increase in the understanding of the yellow silhouetted pedestrian when compared to the flashing orange hand to prompt pedestrians not to begin to cross (78% vs 58%). Although pedestrians better understood the tricolored pedestrian heads than the standard pedestrian heads, the majority of those surveyed did not prefer the new tricolored heads to the standard pedestrian devices.

Observations of pedestrian behavior at crosswalks indicated that the tricolored pedestrian heads did not increase pedestrian compliance at crosswalks. The authors concluded that pedestrians better understood the clearance phase when the tricolored heads are used, however, pedestrians did not show better compliance to these new pedestrian heads than they did to the standard ones. The authors also report no safety benefits in installing the tricolored heads. They further estimate the costs of installation of the new devices to be between $3,000 and $10,000 Canadian per intersection. After weighing these costs against the benefits, the experimenters concluded that the use of the tricolored pedestrian head was not justified.

Another group of researchers examined a second strategy to increase the comprehension of the clearance phase — the use of an LED count down timer that displayed the number of seconds left for the pedestrian to cross (Belanger-Bonneau, Lamothe, Rannou, Joly, Bergeron, Breton, Laberge, Nadeau, & Maug,1994). In this study a pedestrian head that flashed a digital count down of the number of seconds left for pedestrians to cross was compared with a standard pedestrian head. The digital count down pedestrian heads were installed along with a standard pedestrian head with the DON'T WALK indication associated with a steadily illuminated orange hand, the clearance interval associated with a flashing orange hand and a walk phase with a white silhouette of a pedestrian. The digital count down head was the same size as the standard pedestrian head installed along with it. The digital count down lasted 24 seconds, 18 seconds for the walk phase and 6 seconds for the clearance phase. The authors measured pedestrian head turning and vehicle-pedestrians conflicts, and pedestrians utilization at crosswalks were recorded at two experimental and two control intersections in the city of Saint-Laurent, Quebec. A pre- and post-survey questionnaire was also administered to pedestrians at the experimental and control intersections to evaluate the perception of security and safety at the experimental and control intersections as well as their understanding of the pedestrian signals.

A total of 4,244 pedestrians were observed at the experimental and control sites during the pre- and post-phases of the study. A total of 1,918 pedestrians were surveyed during the pre and post phases at the experimental and control sites. The main results of the study indicated that the installation of the digital count down pedestrian head did not increase the pedestrian's understanding of the three phases of the crosswalks, that is the "WALK", "DON'T WALK," and clearance phases. The clearance phase (flashing orange) remained the least understood even with the introduction of the digital count down device. Approximately 80 percent didn't understand the flashing orange; the digital count down device, according to the survey, increased the feeling of safety and security of pedestrians using the crosswalks. This feeling of security was greater for people under 17 years of age or over 65 years of age. These increases in perception of security may actually have a negative impact on pedestrian safety because they may induce pedestrians to engage in less visual searching for turning vehicles because they feel more secure. The digital count down device was associated with a small increase in the level of compliance to the crossing signals at one treatment site and a small decrease at the other treatment site. A decrease in motor vehicle pedestrian conflicts was observed at the treatment site, but a similar reduction in conflicts was also observed at the control site.

The authors did not report on the data they collected on pedestrian observing behavior. On this basis of data reported in this study, the use of the count down pedestrian head was not associated with any increase in pedestrian safety. These finding are consistent with findings discussed by Baass (1990) who reported the results of a study conducted by Druilhe in Toulouse, France, that found no significant change in pedestrian behavior following the installation of a count down pedestrian head. Taken together, the results of these studies show that modifications to the pedestrian signal head designed to increase the understanding of the pedestrian clearance interval at best produce only equivocal improvements in comprehension and no safety benefits. These data also suggest that interventions designed to increase pedestrian or motorist observing behavior are likely to yield greater safety benefits.

4. Use of Flashing Amber Beacons at Unsignalized Crosswalks

One way to alert motorists to the presence of pedestrians in crosswalks not controlled by full traffic signal is to use pedestrian-activated flashing yellow beacons (Bowman, 1995). Van Winkle (1997) described the use of pedestrian-activated beacons at midblock crosswalk locations but did not provide evidence of effects on motorist yielding behavior, pedestrian/motor vehicle conflicts, or pedestrian crashes.

For more than a decade, many Canadian jurisdictions have employed flashing yellow beacons at busy unsignalized crosswalks with a multilane approach. For example, the Halifax Regional Municipality has over 100 of these devices in place. Although no research has been conducted to evaluate the signal effects on motorist yielding behavior and pedestrian conflicts, several studies have examined variables that influence the safety and efficacy of pedestrian- activated flashing beacons.

One way to increase the effectiveness of flashing beacons is to pair them with the pedestrian symbol normally used to indicate a crosswalk (a pictograph of a walking pedestrian). Figure 1 depicts a commercially available pedestrian-activated beacon in common use in Canada. It includes the pedestrian symbol and illuminates the crosswalk at night. Another way to increase the efficacy of these signals is to erect a "YIELD WHEN FLASHING" sign that includes the pedestrian symbol and an amber beacon starburst symbol, posted at a location that would accommodate the necessary stopping distance required to yield for a pedestrian. A photograph showing the implementation of this option is shown at the bottom of figure 4. Both of these interventions increase the continuity of signing features and might be expected to alert motorists to look for pedestrians when the flashing beacons are activated.

Van Houten, Healey, Malenfant, and Retting (1998) examined the effects of these two interventions employed alone and together at two crosswalks using a counterbalanced multiple baseline design. Observers scored whether the pedestrian activated the flashing beacons, the yielding behavior of drivers, and motor vehicle/pedestrian conflicts. Following a baseline condition at both crosswalk sites during which pedestrians-activated beacons that did not include the pedestrian symbol were employed, flashing beacons with the pedestrian pictograph were first introduced at Wyse and Faulkner Streets. Next the "STOP WHEN FLASHING" signs were erected at this site. The two experimental conditions were introduced in the reverse order at the second crosswalk. At both crosswalks "ALERT MOTORISTS," "PRESS BUTTON BEFORE CROSSING" signs were erected on the median strip, and these signs were associated with a sustained increase in the percentage of pedestrians activating the beacon.

Figure 4b shows a photograph on the RA 5 beacon with the pedestrian symbol, and figure 4a shows a photograph of the "STOP WHEN FLASHING" sign.

The percentage of motorists yielding to pedestrians when the beacons were activated during the baseline condition averaged 67.6 percent at Wyse and Faulkner and 67.5 percent at Wyse and Sportsplex. The modification of the pedestrian signal to include a pictograph of a pedestrian increased the percentage yielding at Wyse an Faulkner to 78.0 percent, and the introduction "STOP WHEN FLASHING" sign at Wyse and Sportsplex was associated with an increase in yielding to 76.3 percent. The introduction of both interventions at each site was associated with respective increases to 86.7 percent and 87.1 percent. These results were found to be statistically significant.

Figure 4. The bottom portion (b) of this figure shows a photograph of the RA5 beacon with the pedestrian symbol, and the top portion (a) shows the "YIELD WHEN FLASHING" sign and the overall view of the crosswalk.

The number of conflicts recorded each session when the flashing beacons were activated averaged 1.0 per session at Wyse and Faulkner and 3.0 per session at Wyse and Sportsplex during the baseline condition. The introduction of the modified signal at Wyse and Faulkner was associated with a small decline in the number of conflicts to 0.91 per session, but the introduction of the "STOP WHEN FLASHING" sign at Wyse and Sportsplex was associated with a marked reductions in conflicts to 0.37 per session. The addition of the "STOP WHEN FLASHING" sign at Wyse and Faulkner was associated with a marked decline in conflicts to 0.25 per session, and the introduction of the modification to the pedestrian signal at Wyse and Sportsplex was associated with a small increase in conflicts to 0.67 per session. The percentage of pedestrians activating the flashing beacon remained relatively constant across this experiment, averaging   60 percent at Wyse and Faulkner and 71 percent Wyse and Sportsplex.

The results of this experiment demonstrated (1) that adding the pedestrian symbol next to the flashing beacons or adding a sign prompting motorist to stop when the amber beacons are flashing are both effective in increasing the percentage of drivers yielding to pedestrians when the flashing beacons are activated; (2) that the combination of both of the above mentioned interventions is more effective in increasing driver yielding to pedestrians than either used alone; and (3) that conflicts were only reduced by the sign prompting motorists to stop when the amber beacons are flashing.

Additional research needs to be conducted to determine the best way to employ pedestrian-activated signals at crosswalks. Because the purpose of the flashing beacons is to alert motorists to look for pedestrians in the crosswalk and yield when they are present, a more effective strategy might be to mount animated yellow LED eyes that look from side to side just above the pedestrian symbol. This signal should be as conspicuous as flashing beacons and has the added advantage that it specifically prompts the motorist to visually scan for the presence of pedestrians.

5. Research on the Use of Advance Stop Lines

Another intervention that has been documented to reduce conflicts at crosswalks on multilane roads is the use of an advanced stop bar to encourage motorists to yield farther back from the crosswalk (Van Houten, 1988; Van Houten & Malenfant, 1992). When a motorist stops too close to the crosswalk when yielding to pedestrians, their vehicle can obscure the view of drivers travelling in adjacent lanes that the pedestrian needs to cross next. This effect is greatest when the pedestrian is of shorter stature or when the stopped vehicle is a truck, mini van, or large utility vehicle. On the other hand when motorists stop farther back from the crosswalk, drivers in adjacent lanes and pedestrians have improved sight distance. The greater the distance a yielding vehicle stops behind the crosswalk, the farther away motorists and pedestrians in adjacent lanes can see each other and take appropriate action to avoid a crash. Small increases in stopping distance are associated with large increases in sight distance because sight distance is a related to the arc tangent of the distance stopped behind the crosswalk divided by the distance that needs to be covered by the pedestrian before he or she is clear of the stopped vehicle. Another advantage of advance stop lines is that they can help reduce the probability of a "billiard ball" collisions that could result when another motorists has a rear-end crash with a motorist stopped for a pedestrian. The striking vehicle can rear-end and push the stopped vehicle into the pedestrian.

Van Houten and Malenfant (1992) evaluated the effects of signs reading "STOP HERE FOR PEDESTRIANS" alone and in conjunction with advance stop lines on motor vehicle/ pedestrian conflicts at two experimental intersections equipped with pedestrian activated flashing beacons. Figure 5 shows how the distribution of stopping distances is influenced by the sign alone and the sign plus stop bars. These results indicated that the "STOP HERE FOR PEDESTRIANS" sign placed 15.25 m (50 ft) before each side of a crosswalks traversing a multilane highway can increase the distance that motorists stop behind the crosswalks and that the effects persisted over time. This is also true of the sign plus advance stop bars. Figure 6 shows the use of an advance stop line with a sign.

Data on vehicle/pedestrian conflicts indicated that the sign alone reduced conflicts involving the driver or pedestrian taking evasive action by 67 percent. The addition of the advance stop line reduced this type of conflict by 90 percent compared to baseline levels. These reductions were sustained at 1-year follow up.

The overall effectiveness of pedestrian-activated flashing beacons remains to be evaluated, but it is clear that their use is associated with an increase in the percentage of motorists yielding to pedestrians. When they are used in conjunction with several other treatments such as advance stop lines and warning signs erected at the dilemma zone, they are associated with decreases in motor vehicle/pedestrian conflicts. Evaluation of the crash prevention effects of these pedestrian- activated beacons remains to be done.

6. Research on Interventions Designed to Increase the Conspicuity of Crosswalks

De Guise and Paquette (1990) evaluated the effects of replacing marked crosswalks with yellow colored concrete crosswalks at one crosswalk in Cap Rouge, a small municipality near the city of Quebec. A total of 2,591 observations were recorded in the experimental site and 1,922 in the control site during the pre intervention phase. A total of 3,934 and 2,677 observations were recorded in the experimental and control sites, respectively.

The intervention consisted in replacing the marked crosswalk with yellow colored concrete crosswalk to test four hypotheses:

1. A colored concrete crosswalk will reduce pedestrian delay at the crosswalks compared with a marked crosswalk.

2. A colored concrete crosswalk will increase driver compliance of the crosswalks.

3. A colored concrete crosswalk will increase the comfort level and security of pedestrians particularly children and the elderly.

4. A colored concrete crosswalk will reduce the incidence of delinquent crossing by pedestrians.

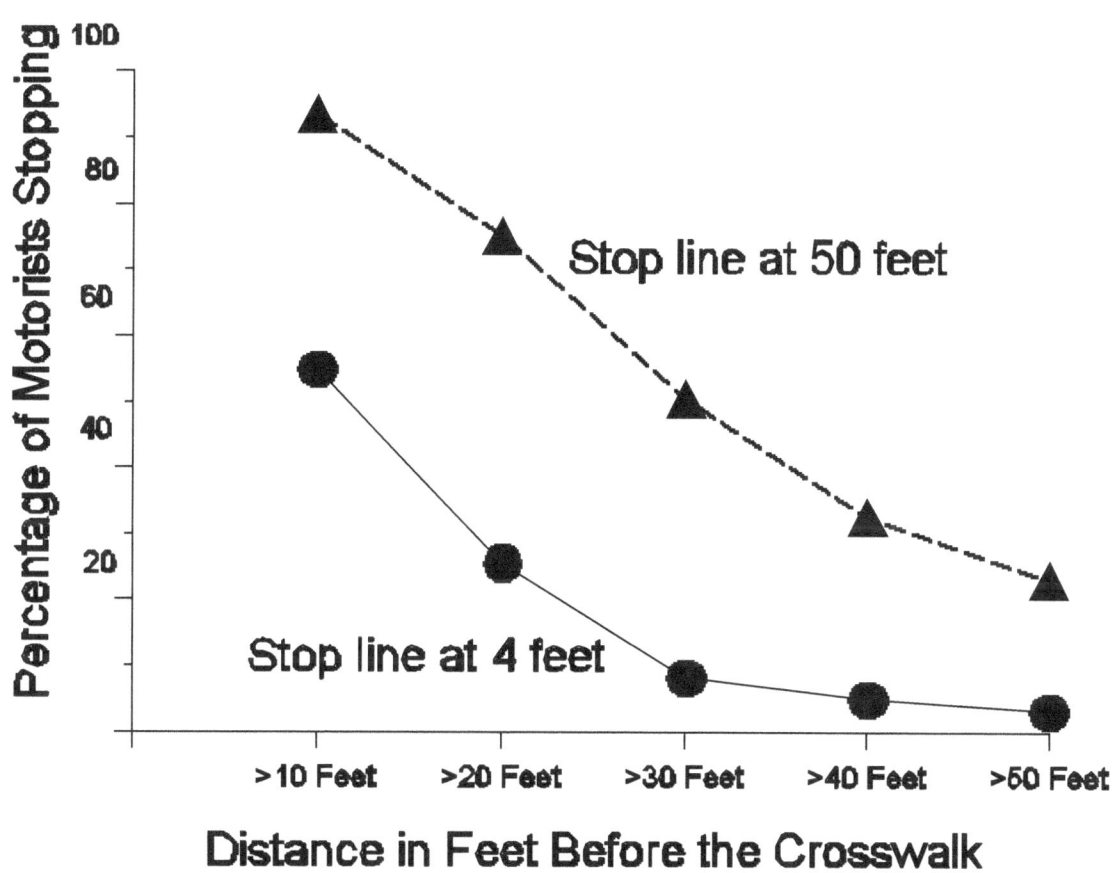

Figure 5. The number of vehicles stopping more than 3.05 m (10 ft), 6.1 m (20 ft), 9.15 m (30 ft), 12.2 m (40 ft), and 15.25 m (50 ft) from the crosswalk during each condition of the experiment at one of the sites.

Figure 6. A picture of an intersection with an advance stop line and a sign prompting motorists to stop for pedestrians at the stop line.

An analysis of the results led to the rejection of all four hypotheses despite the fact that the concrete crosswalk was slightly wider than the existing marked crosswalk and was moved and adjusted to better coincide with the sidewalk. An interesting development in this study is that the colored concrete deteriorated during the winter months of 1989 and the first months of 1990 and had to be asphalted in the spring. The crosswalk was then painted in yellow, and once again all of the four hypotheses were rejected. This study seems to indicate that the conspicuity of the crosswalk may not be a major factor influencing driver yielding behavior or pedestrian crossing behavior at crosswalks.

7. Community Pedestrian Safety Programs in Canada

A complimentary strategy to increase pedestrian safety at signalized intersections is to employ a media campaign aimed on increasing driver yielding behavior. Koenig (1994) reported the effects of a media campaign designed to increase the percentage of left-turning vehicles yielding to pedestrians in 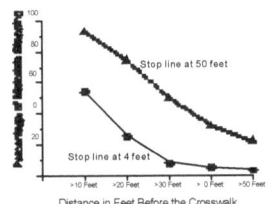 Victoria, British Columbia. They found that the campaign produced a long-term increase in driver yielding behavior at five monitored signalized intersections. A multifaceted program that has been applied in three Canadian provinces is the Courtesy Promotes Safety Program reported by Malenfant and Van Houten (1989). This program consists of education, engineering, and enforcement components that are all implemented together. The educational components included:

1. Flyers sent to each household in the targeted community along with utility bills. The flyers provide safety tips for pedestrians and motorists and address some of the common causes of pedestrian crashes and how to avoid them.

2. Large highway signs erected at locations where they would attract the most attention and provide feedback on the percentage of drivers yielding to pedestrians during the past week along with the record. A photograph of one of these signs is shown in figure 7. The numbers on these signs were changed on a weekly basis and in some communities were sponsored by a corporate sponsor.

3. Small signs were erected at a number of crosswalks instructing pedestrians how to safely cross the street. These signs instructed pedestrians to extend their arm while placing one foot in the street, wait until cars stop, and thank drivers with a wave and a smile. At other sites, the message "EXTEND ARM TO CROSS" was painted in the crosswalk facing the curb.

4. A classroom intervention was designed for all elementary and junior high classrooms. A special folder included a summary of the program, an "I YIELD TO PEDESTRIANS" bumper sticker, a copy of an information pamphlet for each pupil to take home, and a 20-minute lesson plan explaining the proper way to cross the street was prepared for each home room in the target community. The lesson plan taught safe crossing skills by demonstration, role playing, and practice with feedback. Posters explaining the correct way to cross the street were sent to senior high schools and senior citizen homes.

5. A special program was prepared for crosswalk guards. Crosswalk guards received a 2-hour training session and a large supply of pins to give to pupils when they exhibited proper crossing behavior. Although children were encouraged to signal their intention to cross the street by extending their arms, the crossing guard also crossed with the children using a stop sign in the usual manner.

Several of the program components also involved police enforcement. A warning flyer was prepared that contained information on the number of children and adults struck in crosswalks each year as well as the human and financial cost of these crashes. Police conducted many well publicized enforcement operations in each city. These operations involved at least two police officers and a civilian employee, usually a university student who served as a civilian employee to increase the opportunity for the police to stop and educate motorists. The two police officers positioned themselves 91.4 m (100 yd) on each side of the crosswalk. The civilian employee would cross the street whenever no other pedestrians were present to increase the opportunity for police to enforce the law. The civilian employee always placed one foot in the street and extended his or her arm to encourage vehicles to yield and always waved and smiled to thank drivers that yielded.

Whenever a motorist failed to yield to a pedestrian, one of the police officers would pull him or her over and inform him or her that he/she failed to yield to a pedestrian in a crosswalk. The police officer then asked the motorist to produce their drivers license and were given an information flyer. The motorist was encouraged to read the flyer while the police officer filled

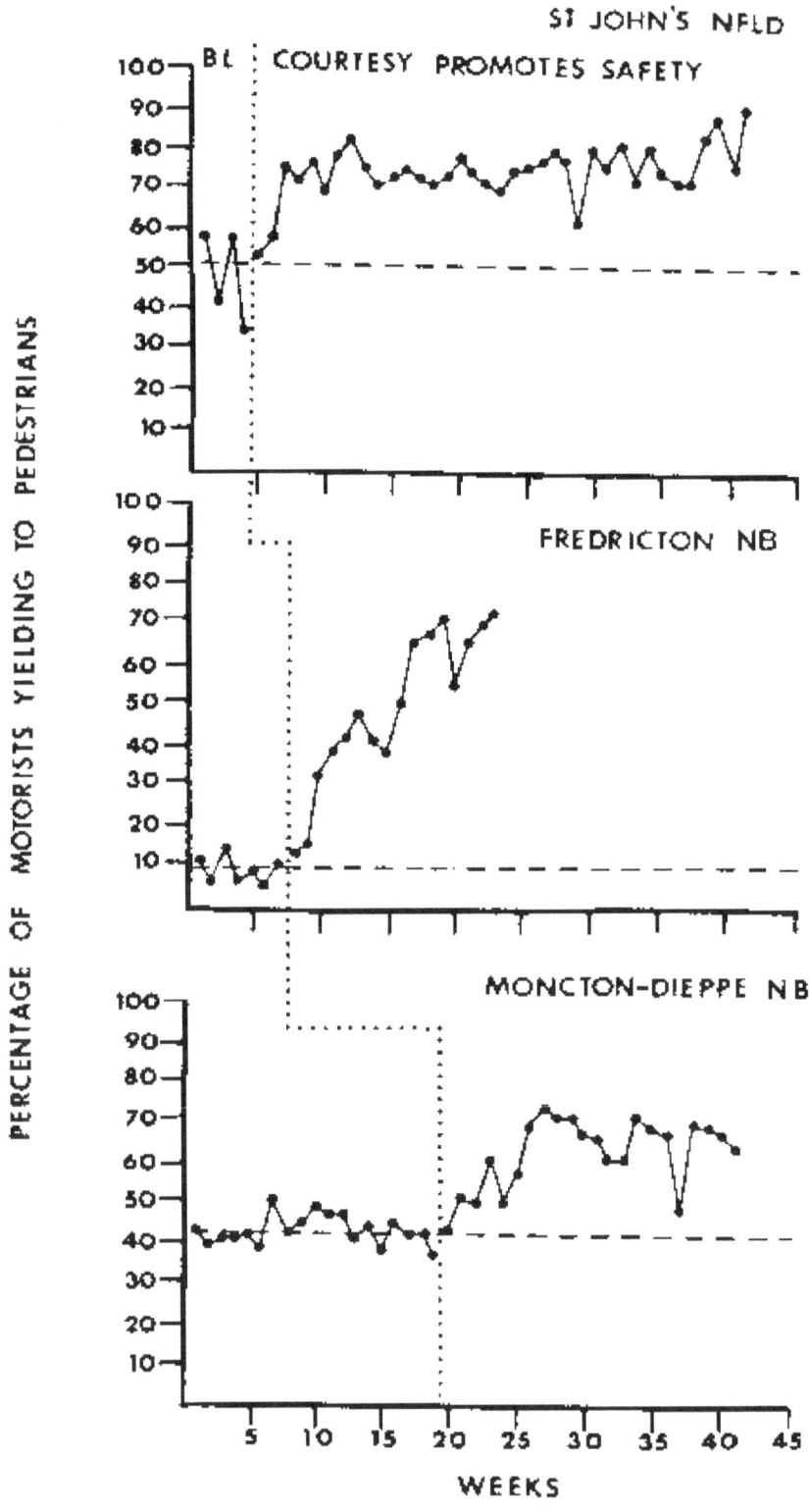

Figure 7. The percentage of motorists yielding to pedestrians in three
Canadian cities before and after the Courtesy Promotes
Safety crosswalk program was introduced.

out a short warning ticket. The officers than gave the motorist the warning ticket and asked him or her to help make their community a safer place to live. The police conducted the enforcement program for 5 hours between 9 a.m. and 4 p.m. during weekdays moving from one crosswalk to another. Police were instructed to spend most of their time at busy crosswalks. This special program was carried out Monday through Friday for the first 2 weeks, and on three randomly selected days during the following 2 weeks, and on one or two randomly selected days during the next 2 weeks. Police also gave pens with the message "Caught Being Courteous" and the name of the police force to some pedestrians that yielded to pedestrians.

The traffic engineering intervention was the use of advance stop lines at a number of busy crosswalks. The advance stop lines were placed 15.25 m (50 ft) ahead of the crosswalk and were marked with "STOP HERE FOR PEDESTRIAN" signs with an arrow pointing down to the stop bars. The purpose of these signs on multilane roads was to increase motorist and pedestrian sight distance by reducing the screening effect of vehicles that might stop too close to the crosswalk.

The Courtesy Promotes Safety Program was implemented and evaluated in three Canadian cities (Malenfant & Van Houten, 1989). The percentage of motorist yielding to pedestrians was evaluated at a number of sites in each city by trained observers. Only warranted crosswalks that were considered problematic because of an excessive number of pedestrian crashes or complaints were included for observation. The results of this experiment is presented in figure 8. During the baseline or pretreatment condition, yielding behavior averaged 54 percent in St. John's Newfoundland, 44 percent in Moncton-Dieppe, and 9 percent in Fredericton, New Brunswick. Data were collected 40 weeks after the program was implemented in St. John's, 23 weeks after the program was implemented in Fredericton, and 25 weeks after the program was implemented in Moncton-Dieppe. The percentage of motorists yielding to pedestrians during the last 4 weeks of the program averaged 81 percent in St. John's, 68 percent in Fredericton, and 71 percent in Moncton Dieppe. Increases in yielding behavior were also associated with a 50-percent reduction in the percentage of pedestrians injured in crosswalks.

One factor that may have potentiated the effectiveness of the Courtesy Promotes Safety Program was the simultaneous implementation of many components designed to improve pedestrian safety. It is likely that the concurrent implementation of many components focuses the attention of motorists and pedestrians on pedestrian issues and has a general synergistic effect. Another factor that should be examined is the impact of adding a media campaign to the package.

Future research should examine how to enhance the efficacy of community intervention programs designed to make it safer and easier for pedestrians to cross a street. Such research could address several interventions to increase the safety of pedestrians at signalized intersections including:

1. The use of a lead pedestrian interval which give pedestrians a 3 or 4 seconds lead while vehicles are held in the all-red condition (Van Houten, Retting, Van Houten, and Malenfant, in press).

2. The use of eyes as part of the WALK indication to prompt motorists to look for turning vehicles.

3. The use of signs to prompt motorists to look for pedestrians (Abdulsattar, Tarawneh, & McCoy, 1996; Zegeer, Cynecki, & Opiela, 1984).

Summary

The two goals of Canadian research in pedestrian safety have been to increase the safety of pedestrians using crosswalks and to make it easier for pedestrians to cross streets. Safety related interventions have focused on prompting pedestrians to look for turning vehicles; prompting drivers to look for pedestrians in crosswalks; the modification of the pedestrian clearance signal by adding a countdown display; the use of advance stop lines to increase sight distance at midblock crosswalks; and increasing the conspicuity of crosswalks. This research has produced mixed results. Prompting pedestrians to look for turning vehicles with signs, pavement markings, or adding animated eyes to the pedestrian signal have all been documented to reduce conflicts between motor vehicles and pedestrians while the addition of a countdown timer for the
clearance interval has not been associated with safety benefits. In regards to pavement markings, the addition of advance stop lines has produced a reduction in motor vehicle/ pedestrian conflicts while increasing the conspicuity of crosswalks has not.

Treatments designed to make it easier to cross the street have focused on: the use of pedestrian-activated flashing beacons at midblock crosswalks and at crosswalks on major roads at intersections not controlled by traffic signals; and the use of multifaceted programs that focus on engineering, enforcement, and education interventions to increase yielding to pedestrians in crosswalks. Although the use of pedestrian-activated beacons have made it easier for pedestrians to cross the street, and are readily used by pedestrians in Canada, the safety value of this intervention has not been clearly demonstrated. However, several studies have shown that the use of special signs and markings may make crosswalks with pedestrian-activated beacons safer. Research also indicates that multifaceted pedestrian safety programs can change community safety culture by modifying the behavior of drivers and pedestrians.

References

Abdulsattar, H.N., Tarawneh, M.S., & McCoy, P.T. (1996). "Effect of "TURNING TRAFFIC MUST YIELD TO PEDESTRIANS" Sign on Vehicle-Pedestrian Conflicts." Paper presented at the 75th Annual Meeting of the Transportation Research Board.

Baass, K.G. (1990). Les Feux de cirulation et la securite des pietons. Texte presente au congres annuel de L'Association Quebecoise des Transports et des Routes. Montreal, P.O.   pp 343-363.

Belanger-Bonneau, H., Lamothe, F., Rannou, A., Joly, M-F.,Bergeron, J., Breton, J.G., Laberge Nadeau, C., Maug, U. (1994). Projet d'experimentation et d'évaluation d'une signalisation numerique pietionnière: Le décompte visuel; Unité de Santé Publique, Hopital du Sacré-Coeur de Montréal, Centre de Recherche sur les Transport, 232 pp.

Bowman, B.L. (1995). *Applications of Supplemental Warning Devices*. ITE Journal, 65, (8), pp 14-19.

De Guise, J. and Paquette, G. (1990). Evaluation d'une traverse pietonnière en béton coloré, Department d'information et de communication, Université Laval, 46 pp.

Druilhe, M. et al. 1987). Pietons: une si lonque attente. Tec No zeg84e 85, September, 1987, pp 36-40.

Gourvil, L., Pellerin, G., Hassan, S. (1994). Evaluation de l'efficasité des feux de pietons tricolorés. 29e congrès annuel de l' A.Q.T.R..(Associations Quebecoise du transport et des Routes Inc.) Valleyfield P.Q. pp 387-406.

Habib, P. (1980). "Pedestrian Safety: The Hazards of Left-Turning Vehicles," *ITE Journal*, 50, (4), pp 33-37.

Jennings, R.D., Burki, M.A., & Onstine, B.W. (1977). "Behavioral Observations and the Pedestrian Accident." *Journal of Safety Research, 9*, pp 27-33.

Koenig, D.J. (1994). "The Impact of a Media Campaign in the Reduction of Risk-Taking Behavior on the Part of Drivers. *Accident Analysis and Prevention. 26*, pp 625-633.

Lane, P.L., McClafferty, M.J. & Nowak, E.S. (1996). "Pedestrians in Real World Collisions." *The Journal of Trauma*, 36, pp 231-236.

Lord, D. (1996). "Analysis of Pedestrian Conflicts with Left-Turning Vehicles." *Transportation Research Record 1538*, pp 61-67.

Quaye, K., Leden, L., & Hauer, E. (1993). "Pedestrian Accidents and Left-Turning Traffic at Signalized Intersections." *AAA Foundation for Traffic Safety*: Washington, DC.

Malenfant, L. and Van Houten, R. (1989). "Increasing the Percentage of Drivers Yielding to Pedestrians in Three Canadian Cities with a Multifaceted Safety Program." *Health Education Research 5*, pp 274-279.

Retting, R.A., Van Houten, R., Malenfant, L., Van Houten, J. & Farmer, C.M. (1996). "Special Signs and Pavement Markings to Improve Pedestrian Safety." *ITE Journal*, 66, (12),   pp 28-35.

Robertson, H.D. (1977). "What is the Message? An Evaluation of Symbolic Pedestrian Signal Displays." Compendium of technical papers of the 47th annual meeting of the Institute of Transportation Engineers, Mexico City, pp 413-422.

Robertson, H.D., & Carter, E.C. (1988). "The Safety, Operation, and Cost Impacts of Pedestrian Indications at Signalized Intersections." *Transportation Research Record* 959, pp 1-7.

Shinar, D. (1978). *Psychology of the Road: The Human Factor in Traffic Safety*, John Wiley & Sons, N.Y.

Van Houten, R. (1988). "The Effects of Advance Stop Lines and Sign Prompts on Pedestrian Safety in Crosswalk on a Multilane Highway." *Journal of Applied Behavior Analysis. 21*, pp 245-251.

Van Houten, R. Healey, K., Malenfant, J.E.L. & Retting, R. (1998) "The Use of Signs and Symbols to Increase the Efficacy of Pedestrian-Activated Flashing Beacons at Crosswalks." Paper presented at the 77th Annual Meeting of the Transportation Research Board, Washington, DC.

Van Houten, R. & Malenfant, J.E.L. (1995). "Increasing Pedestrian Observing Behavior at Signalized Intersections to Reduce the Threat of Turning Vehicles." Paper presented at the pedestrian session at the 74th Annual Meeting of the Transportation Research Board, Washington, DC.

Van Houten, R. & Malenfant, L. (1992). "The Influence of Signs Prompting Motorists to Yield 50 feet (15.5 m) Before Marked Crosswalks on Motor Vehicle-Pedestrian Conflicts at Crosswalks with Pedestrian Activated Flashing Lights." *Accident Analysis and Prevention, 24*, pp 217-225.

Van Houten, R., Malenfant, L. Van Houten, J., & Retting, R.A. (1998). "Auditory Pedestrian Signals Increase Pedestrian Observing Behavior and Reduce Conflicts at a Signalized Intersection." *Transportation Research Record.*

Van Houten, R., Retting, R., Van Houten, J., & Malenfant, J.E.L. (In press). "Field Evaluation of a Leading Pedestrian Interval Signal Phase at Three Urban Intersections." *Transportation Research Record.*

Van Houten, R., Van Houten, J., Malenfant, J.E.L. & Retting, R.A. (1998). "Use of Animation in LED Signals to Improve Pedestrian Safety." Paper presented at TRB.

Van Winkle, J.W. (1997). "Pedestrian-Actuated Crosswalk Flashing Beacons." *ITE Journal, 66*, p 27.

Zegeer, C.V., Cynecki, M.J. & Opiela, K.S. (1984) "Evaluation of Innovative Pedestrian Signalization Alternatives." *Transportation Research Record 959*, pp 7-18.

www.ingramcontent.com/pod-product-compliance
Lightning Source LLC
Chambersburg PA
CBHW081419170526
45166CB00010B/3403